Youth worl
A Student's
your kids. B......... you pass it on to them, read
it for yourself. John Perritt has given us a concise
and direct outlook on technology that's rooted in
the unchanging truths of God's Word. This book
will lead your students into navigating technology
to the glory of God while enabling their own human
flourishing.

Walt Mueller
President, Center for Parent/Youth Understanding
Author of several books, including *The Space
Between: A Parent's Guide to Teenage Development*
and *Youth Culture 101*

As followers of Jesus, we are to do everything for the
glory of God and be thankful for all that He gives us.
But how does that apply to our use of social media?
Or all of the other ways in which technology impacts
our lives? Read this book to find out! It explains that
God's Word gives us wisdom for every aspect of life,
and that Jesus is Lord of all – including technology.

Sharon James
Social Policy Analyst at The Christian Institute
Author of *Gender Ideology: What Do Christians
Need to Know?*

This book offers invaluable help to young people (and older ones) about life in a digital age. The author takes a positive view of technology but encourages us to consider its impact on our spiritual lives. He encourages us to actually *think* about our use of technology and to steer a sensible and God-honouring course past the dangers that are encountered in cyber-space. Subjects include digital addiction, screen-induced loss of sleep, the temptations of pornography and time-management. This book will help readers to be *smart* phone users.

David J. Randall

Author of *Why I Am Not an Atheist* and
Christianity – Is It True?

God's gifts can quickly become idols, and technology is no exception. Along with immense improvements to modern life, it can also become an all-absorbing obsession, distracting us from eternal matters. In nine concise chapters, John Perritt helps young people think through *why* they use technology the way they do, and also why it can be unhealthy and dangerous. Pastors will want parents and youth leaders to make use of this valuable resource.

Charlie Wingard

Associate Professor of Practical Theology and Dean of Students, Reformed Theological Seminary, Jackson, Mississippi

TRACK
CULTURE

JOHN
PERRITT

A STUDENT'S GUIDE TO
TECHNOLOGY

CHRISTIAN
FOCUS

tym

Copyright © John Perritt 2020

paperback ISBN 978-1-5271-0449-5
epub ISBN 978-1-5271-0551-5
mobi ISBN 978-1-5271-0552-2

10 9 8 7 6 5 4 3 2 1

First published in 2020
. by
Christian Focus Publications Ltd,
Geanies House, Fearn, Ross-shire,
IV20 1TW, Great Britain

www.christianfocus.com

with

Reformed Youth Ministries,
1445 Rio Road East
Suite 201D
Charlottesville,
Virginia, 22911

Cover by MOOSE77

Printed by Gutenberg, Malta

CONTENTS

Series Introduction 7

Introduction 9

1. One of God's Greatest Graces 13

2. Technology is Never Bad? 19

3. The God-likeness of Technology 23

4. Every Knee Will Bow ... To Something ... 29

5. Why Am I So Tired? 37

6. Do You Like Me? 43

7. Technology's Dark Side 49

8. Is Mankind Becoming Mindless? 55

9. The Now and the Next 61

Conclusion: There's No Place Like Home 67

Appendix A: What Now? 71

Appendix B: Other Books on this Topic 73

Series Introduction

═══

Christianity is a religion of words, because our God is a God of words. He created through words, calls Himself the Living Word, and wrote a book (filled with words) to communicate to His children. In light of this, pastors and parents should take great efforts to train the next generation to be readers; *Track* is a series designed to do exactly that.

Written for students, the *Track* series addresses a host of topics in three primary areas: Doctrine, Culture, & the Christian Life. *Track's* booklets are theologically rich, yet accessible. They seek to engage and challenge the student without dumbing things down.

One definition of a track reads, *a way that has been formed by someone else's footsteps.* The goal of the *Track* series is to point us to that 'someone else' – Jesus Christ. The One who forged a track to guide His followers. While we cannot follow this track perfectly, by His grace and Spirit He calls us to strive to stay on the path. It is our prayer that this series of booklets would help guide Christ's Church until He returns.

In His Service,

John Perritt
RYM's Director of Resources
Series Editor

Introduction

The pay phone, 8-tracks, Napster, Xanga, MySpace, cassette tapes, the Walkman, compact discs and the cordless telephone – all are casualties of advancements in technology. These bring feelings of nostalgia to some, confusion to others (some of you reading this don't even know what an 8-track is), but humor to most. The list is a collection of technologies that were once breakthrough trends or inventions which now serve as a punchline or are simply forgotten.

Similarly, any book written on technology sets itself up for the same purpose the following year – a joke. The forms of technology mentioned in this book will be like the pay phone to its readers sooner rather than later.

While that may be true, this book seeks to be as timeless as possible. Therefore, there will be forms of technology discussed that are relevant now, but there will be greater focus on truths that will remain true centuries from now.

In this short book, I want us to consider the impact of technology on our Christianity. We wrestle with the questions: 'How is technology shaping our Christianity? How is it affecting us spiritually?' This is the most important way to think about technology, but approaching our topic in this way will – I hope – also allow this to be more of a timeless discussion.

Technology, almost by definition, implies change. Next week you will see more technological advances that are impossible for us to address here. As we will see later, even books are a form of technology, although an older form. And while books are vitally important to the Christian life, they are a form of technology that cannot keep pace with the instant culture we live in.

Even so, books have withstood the test of time. While other forms of technology no longer exist, books are still going strong. Books have endured and will continue to endure in the technological frontier. Whether you are reading this in print or in electronic form (or some new technology without a screen, as many believe we one day will), the written word is here to stay.

Therefore, this book has the eternal Scriptures as its foundation, and those aren't going away. This should encourage you since you should be able to take truths from this book and apply them to any sort of technology that comes your way. For the student reading this who, typically, knows more about technology than the parent who's trying to keep up, this book will give you a biblical framework for living in this technological age.

Often discussions (or books) about technology can foster a certain amount of fear for parents and dread from students – fear that our technology is destroying our children and dread that our parents are going to forbid the technology *all* of our friends have. I don't want to be an alarmist, but there has been enough research produced that should at least sober us to issues surrounding teenagers and the popular forms of technology they use.

Serving in student ministry for over fifteen years and being a part of Reformed Youth Ministries (RYM), where I interact with youth workers and students frequently, as well as being a father myself, I've seen a lot that is concerning. I've had students and parents in my office – in tears – over issues related to technology. Therefore, some of what is discussed in the pages ahead should create an appropriate concern, not fear or anxiety, but concern.

The aim of this book is to get at the heart of why we do the things we do with our technology. It seeks to explain why humans – both children and adults – use their technology in unhelpful ways at times. Let me go further and say that the Bible explains our unhelpful and often sinful technological practices. Hopefully, you will see how clarifying God's Word is on this subject.

If you're a teenager apprehensively reading this, hoping that your parents don't pick it up, or wondering if it's going to bash your favorite form of technology, let me calm you down a bit. As Christians, we cannot oppose technology. To say it another way, the Bible encourages technology. That truth is what we will discuss in chapter 1.

1. One of God's Greatest Graces

Late one January, I received a phone call from my wife that rattled me. I was in Nashville (which is about six to seven hours from my home) at RYM's annual Youth Leader Training. It was around 11 p.m. when my phone rang. When I saw that it was my wife, I was a bit puzzled because I assumed she would be asleep. While it was surprising to receive a call from my wife at that hour, the greater shock came when I answered the phone.

'Hello?' I said. No response. 'Ashleigh, are you okay?' This time I got a response, but it was muffled, through tears.

All of this happened in a matter of seconds, but my thoughts were along these lines: *I'm receiving a call from my wife at a strange hour. When I answer, I can hear her muffled cries and she's not responding. I'm hours away from her (and my children). I'm helpless.*

My knee-jerk reaction to all of this was to shake her out of her stupor. 'Listen!' I said, 'You better start talking right now, because I'm hours away from you and if you're in trouble, I'm hanging up now and calling the police.'

That seemed to allow her to gather herself, but the information I received did not help my nerves. She informed me that our third child was having trouble breathing. She was struggling to get air, and it was getting worse. In fact, this had been going on for some time, and she was worried that our daughter was about to stop breathing. I told my wife to call 911 and I would call our pastor who lived nearby and then call her back. We both hung up and made those calls.

Four to five hours later, the entire ordeal was over. My wife and daughter (and pastor and his wife) were all back in their own beds and everything was fine – even though my wife and I didn't really go back to sleep after the incident.

GOD IS ALWAYS GRACIOUS
God graciously spared my daughter's life that night, but He would have still been gracious had He allowed her to pass from this earthly life into the next. It's hard to say that and believe it, but

it is true that God is always gracious. His grace is not contingent upon our circumstances.

However, His grace was not simply manifested through sparing her life, it was also manifested through *technology*. Think back through the technology utilized in the story.

Phones were used. My wife called me, she called 911, and I called our pastor. A total of four phones were used. There was also a specific number (which is also short) that granted my wife almost instant access to medical professionals who are waiting to respond. Not only that, but those paramedics, doctors, and nurses who responded had made use of numerous technologies to become the professionals who responded that night. In light of that, there's no telling how many technological tools were employed to assist my daughter. God's grace was manifested through all of this – praise God for technology!

TECHNOLOGY IS THE DEVIL

Often, Christian conversations about technology could be accurately described as 'anti-technology.' As we will discuss later, there should be concern over some technology, and Christians must be discerning about specific technologies they use. That said, technology isn't

15

the devil. Not only is it often a way in which we should see God's grace at work, but Christians are actually commanded by God to advance as a species, and this includes technology.

When we look at Genesis 1:26-28, we read:

Then God said, 'Let us make man in our image, after our likeness. And let them have dominion over the fish of the sea and over the birds of the heavens and over the livestock and over all the earth and over every creeping thing that creeps on the earth.' So God created man in his own image, in the image of God he created him; male and female he created them. And God blessed them. And God said to them, 'Be fruitful and multiply and fill the earth and subdue it, and have dominion over the fish of the sea and over the birds of the heavens and over every living thing that moves on the earth.'[1]

There is much to draw from these verses, but one thing we learn from this passage is something theologians refer to as the *creation mandate*. Basically, this is God's instruction to human beings to exercise rule and dominion over the

---

1   We will return to these verses throughout this book – they are foundational!

earth. This does not mean in a harsh, unloving manner. Rather, this describes a gracious rule that strives to care for God's creation.

Part of humanity fulfilling this creation mandate is through technological advances. Think back to the story of my daughter. Humans – for centuries and centuries – have fulfilled the creation mandate in developing medical and communication technologies. That night, my wife, pastor, his wife, the medical professionals, and I were standing on the shoulders of countless individuals to save my daughter.

Put another way, had humanity ceased to produce technological advancements, my daughter may have died that night. While I will be sure to highlight concerns I have about smartphones, that night I thanked God for them!

Main Point

Technology is a gracious gift from God.

Reflect

- What is one of your favorite forms of technology? How do you see God's grace manifested through it?

- Did this chapter change your perspective on technology? Why or why not?
- Think of ways you can give thanks to God through your daily use of technology.

2. Technology is Never Bad?

When I was about fourteen years old, my older sister confessed something to me that made me angry. We both laugh about it now, but in the moment she shared it, I did not laugh.

My sister told me that when she was about five, she colored all over her dollhouse with crayon or marker – I can't remember which. She was old enough to know that she shouldn't do this, and she was also old enough to know she was going to get in trouble. So, she was faced with a dilemma. That dilemma? *How am I going to get out of this mess I've created for myself?* The solution – her two-year-old brother.

My dear sister (we do love each other to this day) decided that she would put me in front of the dollhouse with the instrument of coloring she used. She then went and got my parents and told them that I did this. And since

I was two – AND COULDN'T TALK TO DEFEND MYSELF – I got punished for the *horrific* vandalism she performed on a dollhouse. While I do not seek vengeance by telling this story, and I have forgiven my sister, there is something satisfying about getting it in print.

We all have a sense of justice, so it's good to know the record has been set straight. The focus of the wrongdoing that occurred all those years ago was misplaced.

WHO'S TO BLAME?

In the last chapter I talked about how technology can often get a bad rap and be referred to as an evil in our culture. However, unless you skipped over the last chapter or weren't paying attention, you'll know that we saw technology is a grace from God and something He commands humans to participate in.

Not only is technology a great gift from God, it's also not an exaggeration to say that technology itself is never bad. While this may seem somewhat puzzling when we think of the variety of technologies out there, just think back to the story of my sister and me. The blame was placed on me, but I did no wrong.

Well, we can often place blame on the technology, but the technology didn't do any wrong. You see, at its core, technology is simply performing a function that it was created to do. Who created it? Mankind.

Therefore, we can say that technology is never the guilty culprit; humanity is. Of course, technology can break and malfunction because it's created in a fallen world, but it's the product of sinful humans. So, the blame of 'evil' technologies ultimately falls to you and to me.

Let me borrow an example from the apostle Paul. One of the most misquoted verses in the Bible is, 'Money is the root of all evil.' But the Bible never says that! What it does say is, 'For the *love* of money is a root of all kinds of evils' (1 Tim. 6:10a, emphasis mine).

You see, Paul (and the Bible, for that matter) would be wrong to say that money is evil for a few reasons. First, money can be a great good and grace from God. Money supports churches and missionaries to carry the gospel across the world. Second, money can be used to alleviate sickness and suffering in some cases. To be sure, there are more reasons, but let me continue with the image of the initial story of this chapter (sorry, sis).

Paul would be placing blame on money, when the blame is rightly to be placed on you and me. Very practically speaking, money can never and will never harm anyone. If you took a specific form of currency (wherever you live) out of your pocket and placed it on a table, it would never harm anyone. It would sit there and collect dust. However, when money gets in your hand and in mine, that's when it becomes evil. When we love money more than God or others, it becomes evil, but it is only evil because of us. The same goes with our technology.

Main Point

Technology is not bad, humans are.

Reflect

- Did this change your perspective on technology? Why or why not?
- Can you think of ways in which technology does seem evil? What role did sinful humans play?
- What are some ways you abuse technology? How is your sin impacting a good technology?

3. The God-likeness of Technology

Parts of the Bible have become familiar to much of the world, even people outside the church. One of those sections is the historical depiction of the Israelites' delivery from slavery under the Egyptians in the book of Exodus. Although Ridley Scott's laughably inaccurate retelling of this historical event was filled with errors, the story of the Exodus still found its way into popular culture.

An interesting aspect of the Exodus has to do with the plagues God sends to free His people. As we read about the horrors that were unleashed, we may simply think that the frogs, gnats, darkness, and the River Nile turned to blood, among other things, were simply random ways in which God's power was manifested.

However, many scholars have pointed out the historical context of this event, and how every plague correlated to a false god that the Egyptians worshiped. In essence, the one true God of the Bible was mocking the false gods and displaying His power as supreme.

THE GOD OF TECHNOLOGY

Well, the God who delivered the Israelites from captivity is the same God today. He has been around since before creation, and He is still faithfully protecting His people, and He will carry them home. Unfortunately, humanity is still in the same sinful state, still making false gods out of anything and everything.

As we've said, God is the ultimate Creator of any technology that was or is or is to come. He's the true God of technology, but as we saw in the last chapter, our sinful hearts can make technology evil. Broadly speaking, our sin turns technology into a god.

Many of us know that we can turn anything – even good things – into a false god, or idol, but technology seems to add a deeper layer to this. In one way, our technology can become an idol like anything else, but in another way, it seems to possess characteristics that elevate it to another level.

Consider with me technology's god-like characteristics…

- *Our Technology is Omnipresent* – One aspect of God's being is that He is everywhere. The fancy word for this is omnipresence. Well, think about the Internet for a minute. If I were to ask, *Where is the Internet?* What would you say? I think 'everywhere' would be a fairly accurate answer. While the Internet is not literally everywhere or omnipresent to the degree that the God of the Bible is, it possesses some elements of what our minds consider omnipresence.

- *Our Technology is Omniscient* – Another aspect of God's being is that He is all-knowing. Again, the fancy word for this is omniscience. God knows absolutely everything. Well, if you have a question about anything, where do you go? The Internet, or more specifically, Google. Think about this: what's a question you could ask Google that it doesn't know the answer to? Once again, Google is not omniscient to the degree God is (see Job 38 for just one example), but it possesses some god-like characteristics.

- *Our Technology is Omnipotent* – Lastly, God is also all-powerful. There is no being that even comes close to the power that the God of the Bible possesses. Yet, our technology does wield a certain level of power. As just one example, think of the power of social media. There have been many cultural elites – actors, musicians, politicians, athletes, etc. – who have been destroyed by the power of social media. Some people have lost their careers because 'normal people' utilized the power of a certain social media platform to destroy them. Social media is limited and laughable in comparison to God's omnipotence, but it does possess a power.

THE HANDHELD GOD

Even though technology is much broader than our smartphones, I'm using this technological device to illustrate its god-like characteristics. Now, let's pause to reflect on what I've just communicated, as well as the conclusion from the previous chapter.

The smartphone is omnipresent, omniscient, and omnipotent – not to the degree God is, but it possesses some of those characteristics. Add to that the sinfulness of mankind's heart. What do you think the outcome will be when we take something that is very god-like and put it in the hands of someone who wants to be God? (Deep down that's the desire of all humanity.)

To take this even further, what do you think the outcome will be when we put this in the hands of very young humans? Humans who are eight years old? Humans who have less wisdom and discernment, but who possess a heart that wants to be God? While the outcome can be a number of things, it's not good. If you are a parent or teenager who thinks I'm overreacting or not giving humanity the benefit of the doubt, let me take one current issue as an example for caution – bullying.

Bullying has always been around (see the story of Joseph, specifically around Gen. 37:12ff). Now, it's being manifested via the Internet and has the more specific name 'cyberbullying.' With just a little bit of searching, it is easy to find story after story after story of lives ruined and lost because of humanity's sin manifested online.

In fact, one specific Google search of 'cyberbullying' yielded 18.5 million results in 0.35 seconds. Sadly, new results will come up tomorrow and the next day until Jesus returns and stops it all. Until then, Christians are given God's Spirit to fight the idolatrous nature of our hearts and ask for His wisdom to use our devices with greater caution and discernment.

Main Point

Our technology is god-like and our sinful hearts worship it as if it is God.

Reflect

- How do you see the god-like characteristics of technology in your life?
- How has technology harmed you or someone you know?
- If you know of someone being bullied, please love them by taking action. The wisest first step is prayer. After that, notify a trusted adult to get involved, for example, a parent, pastor, teacher, or coach.

4. Every Knee Will Bow... to Something

Have you ever used Google maps? It's a helpful tool to get where you want to go, but have you ever used the satellite feature? It's so interesting to switch to this feature and get actual satellite pictures all over the earth. It's also interesting (and maybe a little scary) to get an overview of your own house. It can give you a different perspective on something you are pretty familiar with.

Like Google maps, if you zoom out above Scripture and look at the big story of the Bible, it helps you get a broader perspective on things. If you were to zoom out and look at an overview of the story of Scripture, you would see the broad themes of Creation, Fall, and Redemption.[1] Now, explaining each of these

---

1   Restoration is the fourth and final category in the overview of the Bible's storyline. This will be discussed later. I just didn't want you to think I forgot about it.

in great detail is well beyond the scope of this book. Rest assured, I will flesh these categories out along the way, but just keep them in mind.

Back in Chapter 1 (it seems like so long ago, right?), I referenced Genesis 1:26-28 and explained how foundational these verses are to all of life, specifically to this discussion of technology. From this point, I want to use these verses to help us understand why we do the things we do with our technology. In each chapter, I will consider the themes of Creation, Fall, and Redemption.

If you're confused, don't worry. Keep moving forward, and you'll see how these categories are used and what they have to do with technology.

CREATED TO WORSHIP

G. K. Chesterton once said, 'When man ceases to worship God he does not worship nothing but worships everything.'[2] What Chesterton was getting at was the biblical truth that we are all worshippers. All of humanity – believer and unbeliever – have been created to worship God.

---

2   As you will see, there's much debate over the source of this quote: https://www.chesterton.org/ceases-to-worship/ (last accessed March 2019).

Before sin entered the world, Genesis 1:26-28 tells us, God created us in His image; therefore, it is in the DNA of humanity to give Him worship. Later, in Romans, we read that God's fingerprints are on creation for all to see and His existence is written on the heart of every human (Rom. 1:18ff). As many have said, the question is not, *Will you worship?* Rather it is, *What will you worship?*

Even though sin has now entered creation and poisoned our worship, it has not removed the desire to worship someone or something. For some it is food. For others it is friends. It is sports teams, money, fame, comfort, happiness, family, cars – whatever it is that you feel you cannot live without. For many of you holding this book, it's your smartphone.

WORSHIP VS. ADDICTION

Sin poisons our worship. Sin redirects our worship to everything else besides God. This truth led Dr. Edward T. Welch to say, 'Addictions are ultimately a disorder of worship.'[3] Applying this logic to technology,

---

3   Welch, Edward T. *Addictions: A Banquet in the Grave – Finding Hope in the Power of the Gospel* (P&R Publishing: 2001), XVI.

people are worshipping their technology; they are addicted to their smartphones.

Now, I want to be cautious about throwing a word like 'addiction' around. There is definitely a spectrum with regard to addiction, and I think our minds gravitate towards substance abuse and hardcore drugs when we hear the word addiction. Still, it's an appropriate word to use when talking about technology.

In his book, *Irresistible: The Rise of Addictive Technology and the Business of Keeping Us Hooked*, Adam Alter explains that the word 'addiction' had been used long before we applied it to substance abuse. 'Addiction originally meant a different kind of strong connection,' says Alter, '[I]n Ancient Rome, being addicted meant you had just been sentenced to slavery. If you owed someone money and couldn't repay the debt, a judge would sentence you to addiction.'[4]

Behavioral addictions – like biting your fingernails or touching your phone screen – may be in a different category than addiction to heroin, for example, but they are addictions nonetheless and are *very* powerful... even destructive.

---

4   Alter, Adam. *Irresistible* (Penguin Press: 2017), 29.

In the CNN documentary *Being Thirteen: Inside the Secret World of Teens*, psychologists and psychiatrists affirmed that the language of addiction applied to teens' smartphone use is not too strong.

Steve Jobs – the creator of many of these devices – encouraged caution with their use. This was demonstrated by the fact that he wouldn't let his own children use iPads (their own father's creation). Chris Anderson, who edited the magazine *Wired*, would not allow his children to take phones into the bedroom, leading Alter to ask this question, 'Why are the world's greatest public technocrats also its greatest private technophobes?'[5]

I hope this shows you that this book, like pastors and parents, is not irrational in its concern about technology-related addictions. Not only are the creators of these devices communicating caution, many of them are creating devices to make you more addicted.

Senator Ben Sasse, who also sits on the Armed Services' Cybersecurity subcommittee, explains that some of the most brilliant minds are thinking about how to keep us more addicted to our tech. He cites the Pew Research

---

5   Ibid. 2.

Center's statistic that 'nearly 40% of Americans ages 18 to 29 are online virtually every minute they are awake.'[6] Sasse concludes that there is no doubt we are addicted.[7]

TRUE WORSHIP

Statistics show that we look at our phones every 4.3 minutes.[8] If that's not worship, I don't know what is. However, if that's you, please see that the answer is not – stop worshiping! Remember, you were created to worship. The answer is – start worshiping the One who is worthy of your worship. Because of our sin, we will always be struggling against the idols of our heart, but this means we must engage in the struggle.

This reality is foundational to what follows. Simply put, your addiction to your technology is misplaced worship. This means two primary things. First, you are not alone. Your sin moves you to worship technology like everyone else. Second, your addiction to your technology is actually your heart longing for Jesus. I know it

---

6   Sasse, Ben. *Them: Why We Hate Each Other – And How to Heal* (St. Martin's Press: 2018), 175.

7   Ibid. 175.

8   Reinke, Tony. *12 Ways Your Phone is Changing You* (Crossway: 2017), 16.

might seem crazy – and this is something that we will discuss in more detail – but, every time you bow down to your technology, your heart is actually looking for Jesus.

Main Point

Our addiction to technology is actually misplaced worship.

Reflect

- Is your technology the first thing you look at in the morning and the last thing you look at before you go to bed?
- What are some other ways that you worship your technology?
- The knowledge of this reality and prayer are the first steps to fight your addiction. What are some steps you can take to stop worshiping technology?

5. Why Am I So Tired?

Have you ever tried to consider the rest Adam and Eve had before the Fall? Or the rest we will have in heaven? I'm not talking about the literal act of physically taking a nap. I'm thinking more about the inner rest they had. No stress, no worry, no anxiety.

No tests to take or papers to write. No jolting awake in the middle of the night because you forgot about your homework. No bills to be paid. No annoying alarm clocks shaking us awake. The annoyance of tossing and turning at night wasn't even a concept. Adam and Eve were truly refreshed. They never felt groggy.

Even more, they never felt shame or guilt. Just a deep and perfect rest in their soul!

Not so, anymore. Whoever you are reading this book, your rest is far from perfect and complete. You're running on empty to some degree. You're running from something.

Maybe it's difficult relationships at school? Maybe challenges in the home? Maybe it's fear about your future – the career or college you may pursue?

The fatigue you feel right now is your soul telling you, *Life isn't supposed to be like this.* As we continue to go back to Genesis and think of how we were created, we see that perfect peace and harmony were an aspect of the Garden. An aspect that has been broken by the Fall, but an aspect our souls still know and long for.

DAY OF THE LIVING DEAD

There are many aspects of this life that war with our rest, but let's think about the daily life of most teenagers. Consider a few categories of unrest.

First, there's the mental exhaustion. Most teenagers sit in a classroom all day long and listen to their teachers. This requires mental focus that eventually wears you out. Plus, you're trying to study hard to make the grades.

Second, there's the physical exhaustion. Yes, our brains are part of our bodies, so there's overlap here, but think about the physical activity of most teens. Usually there are some extracurricular activities. There's the simple act of walking around and using our muscles to perform daily tasks. Our physical bodies have limitations that wear down each day and exhaust us.

Third, there's the emotional exhaustion. Not just the exhaustion from all the mental and physical activity, but exhaustion from inner turmoil. Exhaustion from body image, exhaustion from fitting in at school, finding friends. Exhaustion from being bullied or mocked by others.

Being a teenager is exhausting... and I haven't even mentioned spiritual struggles yet. To be sure, all of these struggles are related to spiritual issues – all of life deals with the spiritual. But every teen struggles with something. Surviving one day at school is a major feat. It's draining... especially if you're sleep-deprived.

'Fifty-seven percent more teens were sleep deprived in 2015 than in 1991,' says Jean Twenge in her *Atlantic* article that sent shockwaves through the culture.[1] 'In just the four years from 2012 to 2015, 22 percent more

---

1   Twenge, Jean. *Have Smartphones Destroyed a Generation?* [The Atlantic] Last accessed: March 2019. https://www.theatlantic.com/magazine/archive/2017/09/has-the-smartphone-destroyed-a-generation/534198/ This article lays out some of

teens failed to get seven hours of sleep.'[2] In the article she points out the timing of the lack of sleep and how it correlates to the release of the smartphone. Coincidence? I think not.

Lack of sleep might seem like an acceptable part of life in a fallen world, something one simply pushes through or catches up on later. Maybe so, but consider some of the health consequences linked to sleep deprivation: heart disease, lung disease, kidney disease, appetite suppression, poor weight control, weakened immune functioning, lowered resistance to disease, higher pain sensitivity, slowed reaction times, mood fluctuations, depressed brain functioning, depression, obesity, diabetes, and certain forms of cancer.

THE PORTABLE CLASSROOM

As we consider the exhaustion of today's teenagers, it's important to note how the classroom has been modified. I'm not talking about the physical setup of a classroom or educational techniques. I'm talking about the fact that the classroom has become portable.

---

the research from Twenge's book *iGen* (Atria Books: 2017).

2   Ibid.

Let me get at this through my own experience as a teenager.

I, just like everyone who lived as a teen, would have tough days at school. Days when friends disappointed and bullies made life miserable. On those days, I'm sure, I would long for the sound of that school bell reminding me that I could escape. I could leave my classroom behind and find rest from the inner turmoil in the safety of my home.

In contrast, today's students bring the classroom home with them. They carry their classroom around in their purse or pocket. The beauty queen of the class is still sharing selfies with her peers, and the class clown is delivering his punchlines to the class. While the audience has been modified from the school to the online world of social media, the heart and mind of the teen is still on the clock.

Our devices are creating unrest because their glowing lights actually tell our brains to wake up.[3] But they also create unrest because they are forcing us to always perform, to always be present. Rare is the teenager (or

---

3 Studies show that the blue lights of your phones cause your glands to stop producing melatonin so your body begins to prepare for the day. Most phones have night-time settings now.

adult) who puts the device down and sits in silence and solitude, reflecting on their own soul and the Savior who grants it rest. While there's much more to discuss, it seems that this is an appropriate way to end the chapter. Why don't you take five minutes to be still and rest in all that Jesus has done for His children?

Main Point

You were created for rest.

Reflect

- Do you find it easy to get away from your technology? Do you take breaks to get offline and away from social media?
- How many hours of sleep a night do you think you average? Is your phone next to the bed?
- How does your online presence impact you? What ways does it stress or harm you?

6. Do You Like Me?

I heard a pastor share something one time that continues to resonate with me. He said that every human on the face of the earth is walking around asking one basic question of everyone else. That question?

Do you like me?

It seems so simple that some of you may read that and dismiss it, failing to see the ways it is manifested in your daily life. However, let's press into this a bit by considering some of your daily interactions. Why are you nice to people? Why do you try to be funny? Why do you dress a certain way or make sure your appearance is spotless (or disheveled) in public? Why do you hang around the people you hang around and avoid others?

My point is not to create paranoia in your heart, but could it be that the answer to these questions is because you want people to like you? Maybe? Maybe not.

In the last chapter, we reflected on the peace Adam and Eve experienced before sin. In a related way, think about the approval they had. Adam and Eve were perfectly loved by their Creator. While we don't know how much time elapsed between Genesis 2 and 3, we know that no two humans – this side of Heaven who weren't Jesus – have yet to experience the type of love, acceptance, security, comfort, and joy Adam and Eve had with the God of all creation.

FOLLOW ME TO HAPPINESS

It can often baffle us that Adam and Eve turned away from following God and followed the temptation of the evil one. Instead of following God to a perfect life of unending happiness, they followed a lie, and now we have pain following us wherever we go.

When Adam and Eve chose sin, they became enemies of God. While He still loved them and provided for them (not only on that day through animal skins, but also through His promise of Jesus in Genesis 3:15 for ultimate provision), Adam and Eve became sinners and were separated from Him because God could not dwell with sin. There's a sense in which they became orphans on that day. Orphans, longing for acceptance and approval.

As we know, their sin spread to us, so this desire for longing and approval is in our sinful hearts. This is manifested in many ways, but we see this online through 'likes' and 'followers.' Whether we are children of God or not, our sinful hearts deceive us and make us long for approval. Again, do you like me?

'Posting on social media is rarely ever innocent,' says Donna Freitas. 'You don't post simply because you feel like it .... You post with at least the slight hope – if not the profound one – that everyone will see your post and respond positively.'[1] In her book, Freitas shares a story of a popular sorority beauty queen who wasn't happy because the 'likes' and 'followers' were never enough. Freitas concludes, 'Social media can bring down even the most popular and successful students on campus.'[2]

This example illustrates one of the main concerns for so many today. So much research demonstrates the correlation between depres-

---

1   Freitas, Donna. *The Happiness Effect: How Social Media is Driving a Generation to Appear Perfect at Any Cost* (Oxford Press: 2017), 38.

2   Ibid. 6.

sion and anxiety being fed through social media. Many have said that this generation of teenagers is dealing with record-setting rates of depression. The evidence is pretty compelling, and the truth of God's Word seems to back this up.

LOOKING FOR LOVE IN ALL THE WRONG PLACES

Sin has given you and me this longing for approval. It doesn't matter what new technologies come out or what old technologies fade away, this desire will still be there. And if the desire is there, people will think of ways to feed it. In fact, research is showing us that people are hard at work to fuel your desire for acceptance.

Adam Alter discusses this in depth in his book, *Irresistible*, and Cal Newport builds on it when he focuses on two primary forces driving people to their technology. One of those forces is 'the drive for social approval.'[3] Newport supports this claim by citing many popular tech developers who confess that they are designing tech to feed this drive. He goes on to

---

3   Newport, Cal. *Digital Minimalist: Choosing a Focused Life in a Noisy World* (Penguin Random House: 2019), 17.

say, 'The power of this drive for social approval should not be underestimated,' using the example of the product manager for the team that developed the 'Like' button for Facebook and her own wariness of the 'havoc' it wreaks. [4]

The truth we all need to be reminded of is the fact that we have been created in God's image. Although we are marred by sin, being an image-bearer means we have worth, value, and importance. No amount of 'followers' or 'likes' can take that away.

If you believe in Jesus as your Savior, you are God's child, and no one can snatch you out of His hands (see John 10:28-30). Even though our hearts forget this and go looking for that love through many avenues – social media being one – we must cling to the truth of what it means to be God's child.

My friend Joe Deegan wrote a song entitled *Child of the King*. In all sincerity, this song needs to be the anthem of this generation. A generation that is fallen like every other generation, but a generation that's screaming for love and acceptance through their online lives:

---

4   Ibid. 21.

Oh, my soul, you can stop searching; The love that you seek is here in your Father's smile. Oh, my soul, you can let down your guard. You don't have to impress; he's already made you his child... You're no longer an orphan, you're a child of the King.[5]

Main Point

Our soul is longing for love and acceptance, but it can only be found in Jesus.

Reflect

- Have you noticed a lack of happiness that may be linked to social media?
- Do you often find that you wish you had a bigger online presence?
- Do the truths of 'image-bearer' or 'child of God' have any impact on your soul? Why or why not?

---

5   https://www.rym.org/worship

7. Technology's Dark Side

I once heard a pastor say that sex is the biggest religion in the world. In chapter four, we established that every human being will worship something. In light of that, we can see that much of this world worships sex; therefore, we can understand why this pastor says that sex is the biggest religion.

If you look at the current top twenty-five songs or visit the local theater, chances are they will contain something sexual. The dialogue, a chorus, or a scene will aim to satisfy sexual lusts that are in the hearts of mankind.

Why is that?

Well, it's because God created us as sexual beings. Think back to Genesis when God said, 'Be fruitful and multiply and fill the earth...' (28b); He is talking about sex here. It is important to understand that God is also talking about more than sex. He is talking about multiplying

His image throughout creation which involves more than having children. However, God is also talking about multiplying His image through Adam and Eve's offspring; therefore, God is talking about sex.

What I want you to see is that God spoke about sex in the first chapter of the Bible. He wasn't embarrassed to talk about this. Plus, He invented sex. The truth Christians need to grasp is that sex is a beautiful act that God invented before sin came into the world.

CREATION VS. CREATOR

Now that sin has entered the world, humans worship creation instead of the Creator. Sinful mankind has taken the good gift of sex and turned it into a false god that is worshiped. Much of pop culture has fed this worship by making sex the be all and end all. But this false god – like every other false god – is doomed to fail us because only the one true God of the Bible can bring us ultimate fulfillment.

One example of our idolatry of sex is pornography. Pornography is a horrible evil in numerous ways. Sex was designed for God's glory, but pornography attempts to take sex and exclaim to God, *This is ours, and we will do what we want with it.* It is spitting in God's face and attempting to denounce the Creator of sex.

Since the porn industry is a multi-billion-dollar industry, it is all over the internet, as well as many apps on your smartphone. Even if the app is not explicitly dealing with pornography, often it is available through advertisements that pop up in the app. As so many have stated, you don't have to go looking for porn anymore, it comes looking for you.

Dr. Alvin Cooper pointed out the 'Triple-A Engine of Porn' – accessibility, anonymity, and affordability.[1] That is, porn can be accessed 24/7, you can access it without anyone knowing, and it is (for the most part) completely free. Free to the viewer, not to the individuals who have been abused and trafficked into this industry.

CARING FOR GOD'S GOOD GIFT

The chances are very high that anyone reading this book will either know someone struggling or are themselves struggling with pornography. Most statistics tell us that the number of those struggling with porn are the same inside the church as outside the church. Here are some

---

1   Chester, Tim. *Closing the Window: Steps to Living Porn Free* (InterVarsity Press: 2010), 9.

steps to take as we seek to care for God's good gift of sex.

Christians need to know that pornography impacts both males and females. For the longest time, this seemed to affect only men, but today it is impacting females as well. The first step in dealing with this issue is having parents and pastors open their eyes to the ways in which young men and women are struggling and come alongside both.

Once we grasp the reality of the struggle, we also need to understand its severity, and that's the second step. Future marriages are being ruined because of pornography. Individuals are becoming addicted to it. For students or parents who may be tempted to downplay the powerful allure of pornography, you must realize that this isn't something you can dabble in and control. This isn't simply a rite of passage as some parents assume, *Oh, this is just something all boys go through.* This is a sin that destroys, and we must wake up to its harmful power.

A third step is simply talking about it in the home and church. Since we see from Genesis that we are created to be sexual beings, we will not stop sexual desires in our hearts. If we are in positions of authority over children, it is our responsibility to have these discussions and tell them about God's creation of sex. If we are not talking to children about sex, they will run to the Internet and ask it their questions.

For those students reading this, you must seek out a parent, pastor, or trusted adult to talk to. If you are indulging in pornography, you must reach out because this is a powerful sin that will harm you and others in many ways. Ask God for the strength to open up to someone and to seek help.

Because pornography is so prevalent – it's on our apps, video games, movies – it will be a constant battle. And much of your reluctance to reach out for help is due to the shame you feel. Yes, indulging in porn should bring shame because it's sinful. But the good news is that Jesus came to deliver us from shame and to pay the penalty of our sin.

You see, Satan wants you to remain in your sin by staying silent. Satan knows sin grows by keeping it hidden. Satan wants you to be miserable. Satan knows that healing can come about by taking your sin to the cross and asking Jesus for forgiveness. So if you are one who is hiding this sin in guilt and shame, take the first step toward healing by speaking to the God of the Bible who loves to welcome repentant sinners into His arms.

Main Point

God created sex as a good gift, but pornography is a sinful distortion of it.

Reflect

- Do you find yourself looking at images that might not be explicitly pornographic, but still foster lust in your heart? Have you relaxed your standards of what is sinful/wrong?
- Are you someone who indulges in pornography without any remorse?
- If you are someone who struggles with pornography, seek help from a trusted friend and reflect on the forgiveness God the Father offers through the finished work of His Son.

8. Is Mankind Becoming Mindless?

It's crazy to think that there was a time in history when books and reading were viewed in a negative light. Something that would make you dumber, in a sense. According to several sources, the famous Greek philosopher Socrates was against writing because it would weaken the memory and mind.[1]

In biblical times, we know that many religious leaders were required to have the Pentateuch (the first five books of the Bible) memorized. The oral tradition – information passed along by word of mouth – was the primary way history and knowledge was acquired. In some cases, writing utensils were not invented and, in other cases, they were very scarce. Therefore, the human mind was much more conditioned to retain information.

---

1   Google 'Socrates against writing' and see what comes up…. Yes, I get the irony.

Today, when we use the word 'memory,' we often associate it with computers and how many gigs or terabytes they have. We have 'the cloud' and computers that allow us to write at light-speed compared to earlier generations that used an inkwell and quill. One would be wrong to assert that computers and smartphones are bad, claiming that they make us dumber.

That said, we have become dependent upon them and are using our brain's memory in different ways. Just think of the last time you needed directions. Did you commit them to memory or did you use GPS? Think about phone numbers. How many have you memorized?

MIND VS. MACHINE

Some scientists claim that the human mind is the greatest machine. As image-bearers of God, we would affirm the miracle of the human mind. As we think of what the average mind is capable of, even everyday tasks could be viewed as miraculous. When we consider the most brilliant minds in all the world, that should highlight the infinite nature of the mind of its Creator and how God's knowledge is beyond comprehension. The ways we stand in awe of any earthly intellect should always point us to our omniscient God.

An implication of these amazing human minds is the technology they create. Christians should rejoice and affirm the creation of computers and smartphones. They are amazingly creative devices that display what humans, by God's grace, can accomplish. This is a truth we asserted at the beginning of this booklet.

Nevertheless, Christians must also be reminded that their brains do not belong to themselves. First and foremost, our brains belong to the Lord; therefore, we should steward them in a way that honors Him. We cannot dismiss certain practices that may be harmful to our minds with a dismissive shrug of the shoulders. In light of this conversation, we should be cautious of how technology is impacting our minds because they ultimately belong to the Lord. Yes, there is so much good that can assist our minds, and there is much technology that even assists those with various mental handicaps. However, there is also a great deal that negatively impacts our minds and – to be faithful stewards of the brain God has given us – we must use new technologies with wisdom.

For starters, we all know there are apps that display a lack of stewardship with our minds. There are fun games that do, in fact, steward our brains by engaging them and causing us to think. There are also games that may engage our brains but which require so much time that they end up not helping our brains all that much.

Moving beyond apps, devices that foster reading, like e-readers, seem to shape our minds differently. Neuroscientists have shown that reading from a screen versus reading from actual paper utilizes different parts of the brain. Some even say that our brains seem to skim when they read from a screen but focus more when reading from a piece of paper.

I'm not advocating for Christians to destroy e-readers (I own one, and it comes in handy when I travel), but this is an issue Christians must wrestle with when we consider that our brains don't belong to us. I would assert that one solution for Christians would be reading from both – paper and screens. Research seems to say that you must keep this portion of your brain exercised by reading from paper, so be sure to utilize both.

BLESSINGS AND CURSES

Our technological age doesn't seem to be slowing down, and I doubt smartphones, e-readers, and digital storage are going away anytime soon. Once again, Christians must see technology as a gift from God. Technology brings about many blessings that we can sincerely be thankful to God for, but there are also cautions with our technological devices.

It would be wrong to assert that much of our technology is making us dumber, but it would be equally wrong to say that it has no impact on our intellect. The truth is, younger generations who have grown up never knowing a day without smartphones have brains that are being shaped differently than those of generations without smartphones. In many ways, they are being taught to think and learn differently.

For example, if younger Christians have a question or want to learn something, Google and YouTube are a knee-jerk, reflexive action for them. For older Christians, their minds might not go down that path as quickly, because they grew up without those services. Virtual Reality is already here, and that will bring about a new generation who has never known a day without it, and their minds will be shaped by that new technology. It's not necessarily bad or wrong, it's simply a notable difference between generations.

That said, for whatever new game or app or device that's around the corner, Christians must employ them with discernment and caution. We cannot mindlessly indulge technology, because neither the mind or the technology ultimately belong to us.

Main Point

Since God created our minds, we must consider the ways new technology impacts it.

Reflect

- Have you noticed ways in which certain technologies impact your brain? Are there technologies you should consider limiting in your life now?

- Do you have much committed to memory? Do you try to exercise your brain in various ways?

- Have you considered using older forms of technology – like pen and paper – to use different parts of your brain?

9. The Now and the Next[1]

Many people saw the YouTube video entitled, *Look Up*, by Gary Turk.[2] In the short video, he highlights the concerns many people have about our online lives, concerns about the fact that we are, too often, consumed with our screens and missing out on real-life relationships. In short, we need to put devices down and look up.

While Turk himself doesn't assert that we should get rid of all social media (he has over 120,000 subscribers to his YouTube channel and 32,000 followers on various social media platforms), he recognizes the richness of life off the screen.

Thinking back to the Garden of Eden once again, we know Adam and Eve were created

---

1   If you want to think more about time and priorities for the Christian, please consider my book, *Your Days Are Numbered: A Closer Look at How We Spend Our Time and the Eternity Before Us* (Christian Focus: 2016).

2   Over 600 million according to his site.

to live forever. Eternity was written on their hearts. Though they sinned, this notion of eternity is still present in their hearts and minds, and Turk, whether he believes this or not, is tapping into that reality.

Much of his video highlights the missed moments with friends and family, the notion that time is passing us by. Regret is a main theme of the short video and all of this bumps up against our notion of eternity. You see, we still have souls inside us that will never die. They were not created to function in the confines of time. Therefore, whenever we feel like we don't have enough time, or we feel like we've wasted time, it is echoing this truth of the here and now and the life to come.

THE INSTANT EVERYTHING

Convenience and technology are almost synonymous. When we consider many of the technologies that we use most frequently, convenience is a major factor. Much of our dependence – even addiction at times – is grounded in the factor of convenience. And some of the convenience lies in the fact that you don't have to wait for much anymore.

What's that song on the radio? Your device can listen to it and tell you in seconds. Who won the first Super Bowl? Google just told me in 0.99 seconds. Download apps, songs, movies in a flash. Search for something on your search engine of choice, and it will most likely finish your thought before you can even type it. Don't want to wait on that item from Amazon? Expedite shipping and get it the next day.... Wait, I forgot, there's same-day shipping now.

Christian or not, every human knows that death is a reality. All of our lives will come to an end. In light of this reality, instant seems like a great gift. From a Christian perspective, our time and days don't belong to us; therefore, we need to spend them wisely. Well, technology seems to be assisting us with efficiency.

At the same time, it's also sucking the life out of us. Have you ever gone on YouTube and clicked on a video... then another... then another? Have you picked up your smartphone for a specific reason, but then got sucked in to notifications only to look up later and ask, *What was I looking for, again?* Our technology is designed to attract us and distract us from the lives we live.

Consider the old game, *Flappy Bird*. The game required the participant to tap the screen of the phone to make the bird flap its wings. The player would do this to direct the bird through obstacles and reach a high score. It definitely engaged the mind and required a certain level of hand-eye coordination, but it was also pretty mindless.

At the height of Flappy Bird's popularity, the creator of this game was making $50,000 a day... a day! Why was he making that insane amount of money? Because people wouldn't stop playing. People were spending hours and hours of their lives playing. You would think he would be happy, but he was torn because he knew people were sharing their addiction stories online. The designer shut the game down because he couldn't handle the guilt.[3]

I can remember students talking about Flappy Bird – I had the app for a short time and played it a bit. I can remember people talking about someone's high score, but here's a question for you: *Does anyone care about your Flappy Bird high score today?*

Whether you are playing a pointless app, binge-watching through a streaming service, or scrolling through pics and posts on a social media platform, all these activities take time. Seconds, that add up to minutes, that add up to hours, that add up to days, that add up

---

3   Adam Alter discusses this in his book, *Irresistible.*

to years. Given a recent stat that says teens spend up to nine hours a day on their phones, those days will add up pretty quickly. (I'm not picking on teens; parents are just as guilty, if not more.)

TIME ISN'T ON YOUR SIDE

Since we have eternity etched in our hearts, time will always be at war with that. We always feel rushed or behind on something. We're often plagued with guilt for how we spent our time. At the risk of creating more stress and guilt – I hope not – the Bible does charge us to use the time we have in a wise way.

If you're a Christian, you believe that Jesus Christ lived a perfect life and died an atoning death on the cross. His perfection is given to you, and all your sinful filth was placed on Him. Let me ask you a sobering question: *Do you think He died a bloody death on the cross so you could 'like' more pictures? Scroll endlessly? Tap your screen more? Binge-watch your days away?*

As Christians, we have freedom to enjoy social media and movie consumption, but we must enjoy with caution. Not only do we need to be cautious of the content, we also need to be cautious of the time. Jesus sacrificed a lot to bring you to heaven with Him. While we cannot use our hours and days to earn the amazing gift of eternity He freely offers, shouldn't that good gift motivate us in the way we spend our time? The Bible is pretty clear that it should. Even though knowing how best to spend our time is challenging, God was gracious enough to give us plenty of priorities and valid ways to redeem the time in His Word.

In fact, why don't you put this book down, pick that book up, and see what it has to say.

Main Point

Our days don't belong to us, so let's ask God to give us the strength to use them wisely.

Reflect

- What do you spend the most time on each day? What technology do you spend the most time using?
- Do you feel like there's a certain device you are enslaved to? Can you take steps to cut back the amount of time you use it?
- How much time do you spend reading God's Word? Praying? Showing hospitality? Sharing the gospel with unbelievers? These are wise ways to spend your time.

Conclusion: There's No Place Like Home

Well, we've covered a lot of ground, and although there's so much more to talk about, we need to wrap this up. Hopefully, you're encouraged by the good of technology, and perhaps you see it as a greater gift than you once thought. I also hope this has given you a theological grid to understand and foster deeper wisdom in your use of technology.

As I mentioned, discussions focused on technology can feed fear or anxiety in people because they often move our minds to think about the future. The future can be frightening because it deals with the unknown. *What new invention is around the corner? How will it impact me spiritually? How do I need to prepare for that?*

Thoughts like this can lead us to despair and give us a desire to run away from this digital life and live on an island. Older generations may long for their childhood home in a time when these new technologies didn't seem to be invading their homes as they do now.

The reality is, not only is it impossible to turn back time, but this world is not our home. You see, one of the allures of technology we did not specifically address is the truth that much of our technology is a refuge for us. I know that's not a word we use often, but a refuge is simply a place we run to for safety and comfort.

When your day has been difficult, maybe you run to an online video-streaming service for refuge? You watch video after video to forget about your troubles. Maybe when you walk into a room crowded with people, you reach for your smartphone because it brings a sense of comfort… refuge.

The online world can often be an escape for us. A place that does, indeed, offer refuge, but it's a fleeting refuge. While we can enjoy sincere relief and comfort from the distresses of our world, technology is not meant to be our true refuge.

Ever since Adam and Eve sinned against God, the longing for true refuge has been a desire in the hearts of humanity. At their initial sin, Adam and Eve sought refuge in fig leaves.

Yet, even in their disobedience, God gave them a better refuge in animal skins. It was a refuge that required bloodshed. It was a refuge that pointed them (and us) to the true refuge, Jesus Christ.

So our longing to return to a simpler way of life, our longing to return home, our longings we try to fulfill through our technology are often just longings for Jesus. When we look at the stats that seem to prove we are addicted to our phones or that we spend a certain number of hours a week on our phone – all of it illustrates humanity's longing for Jesus. Longing for the return of our Creator. Longing for the relationship with our God to be restored. Longing for the true home.

Until that day, Christians are to remain faithful. We are to live in the times God has called us to live in. To engage and use technology for the glorifying of His name. To enjoy this home He has given us as we long to dwell with Him in our true home.

Appendix A: What Now?

- Look around the room you're currently sitting in and notice all the technology. Give thanks to God for it.

- Try to make thanksgiving for technology a part of your daily life.

- Add up the hours you spend each day playing video games, video streaming, social media, etc. See if you can cut an hour or two away and invest that somewhere else.

- If you don't do this already, take daily breaks from your smartphone and social media – at meals, first thing in the morning, right before bed, etc.

- Take extended breaks from your smartphone – one day a week, a few weekends out of the year, or an entire week.

- Spend some time outside. This could be sitting in the back yard, going for a walk in your neighborhood or a park, camping, etc. Do something that gets your eyes away from a screen and onto God's creation. God's Word tells us that He speaks to us and teaches us through creation (Ps. 19 & Matt. 6:25-34)

- If you're struggling with pornography, reach out to your parents for help. If you feel like you cannot reach out to them, reach out to an older Christian influence – youth worker, pastor, mentor at church.

- Make daily prayer and Bible reading a part of your day. Ask God to give you this desire. There are so many temptations through smartphones & social media, so you need God's Word and prayer impacting your hearts.

- Get together with some friends and put your phones in another room. Go out to eat with your friends and leave your phones in the car.

- Don't text on the phone while carrying on a conversation with another person. Even though this is becoming normal practice, it's actually not respecting the other person.

Appendix B: Other Books on this Topic

Challies, Tim, *The Next Story: Life and Faith after the Digital Explosion* (Zondervan, 2011).

Detwiler, Craig, *iGods: How Technology Shapes Our Spiritual and Social Lives* (Brazos Press, 2013).

Dyer, John, *From the Garden to the City: The Redeeming and Corrupting Power of Technology* (Kregel Publications, 2011).

Reinke, Tony, *Competing Spectacles: Treasuring Christ in the Media Age* (Crossway, 2019).

Reinke, Tony, *12 Ways Your Phone is Changing You* (Crossway, 2017).

Watch out for other forthcoming books in the
Track series, including:

Sanctification
Technology
Prayer
Body Image
Music
Rest
Addiction
Marketing

LIGON DUNCAN
& JOHN PERRITT

TRACK
DOCTRINE

A STUDENT'S GUIDE TO

SANCTIFICATION

A Student's Guide to Sanctification

LIGON DUNCAN & JOHN PERRITT

Knowing that we have been saved by what Jesus has done rather than by what we have done is amazing. But how does this knowledge affect the way we live? What's the point in being good if we will be forgiven anyway? Actually the Bible says that God's forgiveness frees us to live for Him and through the Holy Spirit we can grow to become more and more like Jesus. Ligon Duncan and John Perritt dive into what that means in this short book.

978-1-5271-0451-8

EDWARD T.
WELCH

A STUDENT'S GUIDE TO
ANXIETY

A Student's Guide to Anxiety

Edward T. Welch

We all know the feeling. That nervous, jittery, tense feeling that tells you that something bad is just ahead. Anxiety can be overwhelming. But the Bible has plenty to say to people who are anxious. This book will help us to take our eyes off our circumstances and fix them on God.

978-1-5271-0450-1

Reformed Youth Ministries (RYM) exists to reach students for Christ and equip them to serve. Passing the faith on to the next generation has been RYM's passion since it began. In 1972 three youth workers who shared a passion for biblical teaching to youth surveyed the landscape of youth ministry conferences. What they found was an emphasis on fun and games, not God's Word. Therefore, they started a conference that focused on the preaching and teaching of God's Word. Over the years RYM has grown beyond conferences into three areas of ministry: conferences, training, and resources.

- **Conferences:** RYM's youth conferences take place in the summer at a variety of locations across the United States and are continuing to expand. We also host